INVESTIGATING MATHEMATICS

GRAPHS

Ed Catherall

CHILDRENS PRESS INTERNATIONAL

Investigating Mathematics

Areas
Graphs
Numbers
Sets

Angles
Shapes
Volumes
Weight

Library of Congress Cataloging in Publication Data
Catherall, Ed
Graphs
(Investigating mathematics/
Ed Catherall) Includes index
Summary: Offers instruction in making and using pictographs, bar graphs, circle or pie graphs, and line graphs.
1. Graphic methods—Juvenile literature.
[1. Graphic methods] I. Title. II. Anstey, David, ill. III. Series: Catherall, Ed.
Investigating mathematics
QA90.C296 1983   001.4'226   82-19886
ISBN 0-516-02281-4

1983 American Edition published by Childrens Press International

First published in 1982 by Wayland Publishers Limited
49 Lansdowne Place, Hove, East Sussex BN3 1HF, England
© Copyright 1982 Wayland Publishers Limited

Illustrated and designed by David Anstey
Typeset by Tunbridge Wells Typesetting Services Ltd.
Printed in Italy by G. Canale & C.S.p.A., Turin
Bound in the U.S.A.

Contents

Chapter 1 Pictographs

Chapter 2 Bar graphs

Chapter 3 Circle graphs or pie graphs

Chapter 4 Line graphs

Chapter 1 Pictographs

Why make graphs?

Ask your friends to lend you a photograph or drawing of each of their pets. Ask them to put their name and some information about their pet on the back of each picture.

Make a list of all the pets owned by your friends. (Picture 1)
Include information about each pet.

Sort the pictures and arrange them in groups on a large sheet of paper. (Picture 2)

How many different kinds of animals do your friends keep as pets?
Which is the most common pet?
How many friends keep dogs?
Does anyone own a black cat?
How many pets have fur?
Which is the most unusual pet?

Was it easier to answer these questions using the list or the pictures? Notice that you can arrange the pictures in different ways.

A graph gives information in the form of a picture. When you arrange the pictures of the pets on a sheet of paper you make a picture graph pictograph.

①
John has a dog	Kate has a dog
Mary has a cat	Sue has a cat
Tom has a goldfish	Ben has a goldfish
Jo has a dog	Tim has a cat
Fred has a tortoise	Pam has a cat

②

Pets

Making pictographs

Stand on a sheet of newspaper.
Draw around the outline of your left shoe.
Cut out the outline of your shoe.

Draw around the outline of the left shoe
of each of your friends.
Cut out each shoe outline.
Put the owner's name and shoe size on
each shoe outline.(Picture 1)

Arrange the shoe outlines in order of size.
Where does your shoe come in this order?

Arrange the shoe outlines as a graph. (Picture 2)
Which shoe size is most common?
Which of your friends has the smallest shoe size?
Do people of the same age have the same shoe size?
Do girls wear smaller shoes than boys?

Draw the outline of your left hand and
the outline of the left hand of each of
your friends.
Cut out each hand outline.
Put the owner's name on each hand outline.
Arrange the hand outlines in order of size.
Where does your hand come in the order?
Do girls have smaller hands than boys?
Is there any relationship between the
size of the hand and the size of the shoe?
Can you make a graph
using the hand outlines?

① John Smith Size 2

②

Shoe size			
3	TOM	KATE	ALAN
2	JOHN	MARY	
1	JANE	TIM	

5

Using symbols

Make a calendar for next week. Mark the days of the week and the dates. (Picture 1)
Record daily the weather on your calendar for each day.

Use an arrow symbol for the wind direction. (Picture 2)
Use a thermometer symbol for the temperature. (Picture 3)
Use symbols to show the types of weather. (Picture 4)
Make up extra symbols to show other weather types.
Put all these symbols in a key at the bottom of your calendar.

What symbols are used on your local television weather maps?

Look at your completed chart.
Which was the hottest day?
Which was the coldest day?
Which was the most common wind direction?
Which day had the most typical weather for this time of year?

If you were to record a month's weather, what different symbols would you need?
What symbol would you use to show wind speed?

①

	Sun Oct 9	Mon Oct 10	Tue Oct 11	Wed Oct 12	Thur Oct 13	Fri Oct 14	Sat Oct 15
Wind							
Temp.							
Type							

② Wind direction

③ Temperature

Hot   Warm   Cold

④ Weather types

Rain

Sun

Clouds

	Sun Oct 9	Mon Oct 10	Tue Oct 11	Wed Oct 12	Thur Oct 13	Fri Oct 14	Sat Oct 15
Wind	↗	→	↓	↙	←	↗	↗
Temp.	🌡	🌡	🌡	🌡	🌡	🌡	🌡
Type	clouds	clouds	sun	sun	sun	rain	rain

Key   Wind direction
Temp.
Type

6

Taking a traffic census

Take a notebook and pencil with you to a busy street.
Stand in a safe place and watch the traffic.
What different kinds of vehicles pass you?
Draw columns in your notebook for each kind of vehicle. (Picture 1)
Record the number of different kinds of vehicles that pass you in one hour.

Return home. Draw on thin cardboard a symbol that represents each kind of vehicle.
Carefully cut out each symbol to make a stencil.

Draw your pictograph on a large sheet of paper.
Put the title, date, time and place at the top of the paper. (Picture 2)

Use your stencils to draw onto your graph the outline of each vehicle that passed you. (Picture 2)
Use crayons to color in each outline.
Use a different color for each kind of vehicle.
How many cars passed by in one hour?
Does your pictograph show anything unusual in the traffic pattern?

①

Traffic census taken at the High St. New Town at Time . . . Date . . .	
Cars	III
Bicycles	II
Trucks	IIII
Buses	II
Taxis	IIII
Motorcycles	III

②

Traffic census taken at the High St. New Town at Time . . . Date . . .	
Cars	
Bicycles	
Trucks	
Buses	
Taxis	
Motorcycles	

7

Coping with large numbers

If you take your traffic census on a busy street, you may have large numbers to put on your pictograph (see page 7).

To save space you can use a code such as one car outline represents ten actual cars. Forty cars would be represented by four cars on your pictograph

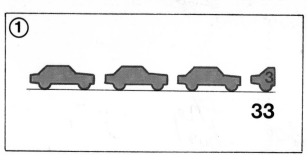

33

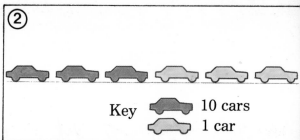

Key   10 cars
      1 car

If you need to represent a number less than ten, use part of the car outline. Here the actual number of cars must be written on the pictograph. (Picture 1)

Another way to represent large numbers is to have one color for ten cars and a different color for one car. So 33 cars would be shown as in Picture 2 The code used must always be included in the key on the pictograph.

Use this color code to draw 44 cars; 21 cars; and 53 cars.
How many car outlines would you need to show 27 cars?

③ World population

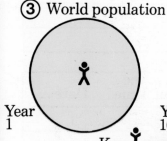

Year
1

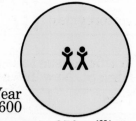

Year
1600

Year
1950

Year
2000

Key 🧍 represents 250 million people

Picture 3 is a pictograph showing the world population growth.
One human outline represents 250 million people.
How many people lived in the year AD 1600?
How many people lived in the year AD 1950?
How many people are estimated to be alive in the year AD 2000?

Chapter 2 Bar graphs

Making bar graphs

Ask your friends for their birthday dates.
Make a graph showing your friends' birthday months. (Picture 1)
Put the months at the bottom of a sheet of paper.

① Birthday months

			Paul								
			Sue					Jane		Mike	
Mary		Tom	Fred	Bill		Gary		Alan		Kate	Anne
Jan	Feb	Mar	Apr	May	Jun	Jul	Aug	Sep	Oct	Nov	Dec

Which month is your birthday? Draw a square above that month.
Draw a square above each month in which a friend has a birthday.
Write each friend's name in a square above their birthday month.
Notice that the columns of squares form bars. You have made a bar graph.
Which is the most popular month for birthdays?
Who has a birthday in your birthday month?

Make a graph for the color of cars.
Stand in a safe place and record the
color of each car that passes.
Which is the most common car color?

Which television show do your
friends like the most?
Record your results as a graph.

Send a copy of your graph to your
local television station.

What other television survey could
you make a graph of?

9

Grouping your facts

What is your weight? Ask each of your friends how much they weigh? Record their weights in a bar graph. If you have a lot of friends you will have a very large graph. Notice that a large graph is difficult to read. (Picture 1). Here is a graph showing the weights of 24 children. All weights are shown in metric measurement. To convert kilograms to pounds remember 1 kilogram equals 2.2 pounds.

① Weight in Kg of my friends

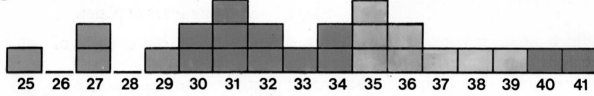

| 25 | 26 | 27 | 28 | 29 | 30 | 31 | 32 | 33 | 34 | 35 | 36 | 37 | 38 | 39 | 40 | 41 |

The graph is much easier to read if you put the weights into 5 groups. (Picture 2)

Make a bar graph showing the weights of your friends in 5 groups. How many 5 groups do you need?

Which group contains the weights of most of your friends?

What weight groups do you need to make this graph easy to read?

②

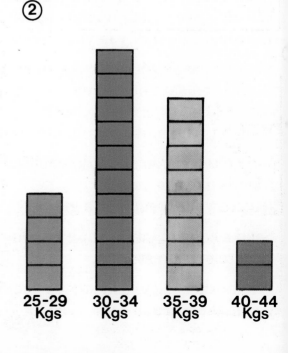

25-29 Kgs   30-34 Kgs   35-39 Kgs   40-44 Kgs

Selecting a suitable scale

How tall are you?
Ask each of your friends their height.
What is the height of your tallest friend?

How could you put this height onto a
sheet of paper?
If you have a sheet of paper as long as
your friend you can draw the graph,
although large graphs are difficult to
read (see page 10).

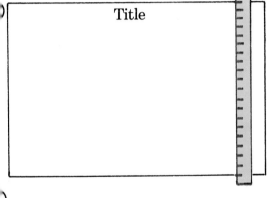

Title

The best way to draw the graph is to
scale down all of your friends so that
they fit onto your sheet of paper.

Measure the height of your sheet of
paper.
Allow space for the graph's title.
How much paper do you have left for your
graph? (Picture 1)

Divide the height of your tallest friend
by the height available for your graph.
This is your **scale**.
If your friend is 60 inches tall and the
paper is 10 inches, then $60 \div 10 = 6$.
So 1 inch of paper represents 6 inches of
your friend's height.
Draw this scale on the left side of your
paper. (Picture 2)
Draw in all your friends' heights. Write
each friend's name on his or her bar.
Draw in all your friends' heights. Write
each friend's name on his or her bar.

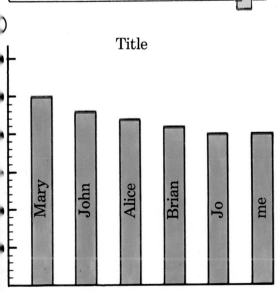

Title

Mary John Alice Brian Jo me

Scale 1 cm = 6 cm
(1 cm = ·394 inches)

Horizontal bar graphs

Borrow a watch that measures seconds.
Ask a friend to time you while you run in a straight line for exactly one minute.
Use paces to measure the distance that you run.
Time your friends as they run for one minute.
Pace the distance run by each friend. ①
Record these distances.
Who ran the greatest distance?

Measure the width of the paper that you will use for your graph. (Picture 1)
Remember to allow for margins.
Divide the longest distance run by the width of your paper. This is your scale (see page 11).
Draw this scale at the bottom of ② your paper. (Picture 2)
Make a graph of the distances run by your friends in one minute.
Put each friend's name on a bar.
You have made a horizontal bar graph.
How many of your friends can run farther than you?

Draw a graph of how far your friends can **walk** in one minute.
Will you use the same scale?

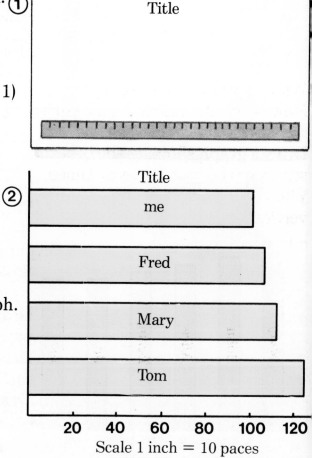

Title

me

Fred

Mary

Tom

20   40   60   80   100   120
Scale 1 inch = 10 paces

Which graph is better?

Bar graphs can be vertical (see page 11) or horizontal (see page 12).
How do you know which one to choose?
If the graph is about height or depth, then a vertical graph is better.
If the graph is about time or distance, then a horizontal graph should be chosen.

Go to your local library.
Find the rainfall for each month over the past year in your area.
Graph this information as a vertical graph.

Find out the distance from your town to other towns.
Graph this information as a horizontal graph.

Which sport do you like best?
How is this sport scored, measured, or timed?
Draw a graph showing the best team scores or the best athletes' times.
Will your graph be a horizontal or a vertical graph?

Comparison bar graphs

Go to your local library. Find out the monthly rainfall for the last two years in your area. Make a bar graph to show the monthly rainfall for each year. Make this graph a comparison graph. (Picture 1)

What can you tell from this graph?

Draw a comparison graph to show the success of your favorite football team over the last two years.

Draw comparison graphs to show the success of your school teams over the last two years.

Comparison of car braking distances. Distances are shown in metric measurement.

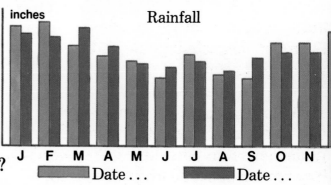

To convert distances to miles remember 1 kilometer equals 0.621 miles

What does this graph tell you?

More difficult comparison graphs

What were the scores of your last three mathematics tests?
Show these scores in a horizontal bar graph. (Picture 1)
Mark your score and the maximum possible score.

Mary scored on her tests:
14 out of 20; 15 out of 20; and 17 out of 20. (Picture 1)

① Mary's test scores

14	20
15	20
17	20

John tried 20 shots at the basket during a basketball game and 17 were successful.
He graphed his result. (Picture 2)
In the next game he managed 8 out of 10 attempts and in the following game he made 18 baskets out of 25 attempts.
Try drawing John's graphs.
Can you compare the results?
Which was his best game?

John converted his scores into percentages.
He divided his successes by the number of attempts
and multiplied each answer by 100.

For example $\frac{17}{20} \times 100 = 0.85 \times 100 = 85\%$

So: 17 shots out of 20 became 85%
8 shots out of 10 became 80%
18 shots out of 25 became 72%
Notice how easily John's scores can now be compared.

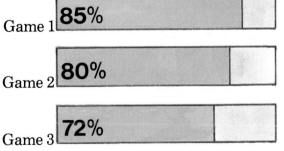

② John's basketball successes

Game 1 | 85%

Game 2 | 80%

Game 3 | 72%

Directional graphs

Where is your nearest commercial airport?
What airlines use this airport?
What kinds of planes do they use?
Where do planes fly to from this airport?

Draw an accurate map showing these air routes. (Picture 1)
Remember to include your distance scale on your map.
Your map is a directional graph of distance.

On a sunny day place a stick upright in the ground.
Mark the position of the shadow every hour on the hour.
What do you notice about the lengths of the shadow?
When is the shadow the shortest?
Why is this?
Draw your shadows to scale. (Picture 2)
You have made a directional graph of your shadows.

Measure the shadows on a different day. What do you notice about the lengths of the shadows? What do you notice about the angles between the shadows?
Why is this?

Scale 1 : 40 million
¼ inch = 100 miles

①

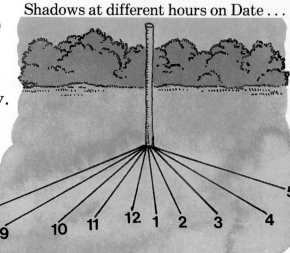

Shadows at different hours on Date . . .

②

8   9   10   11   12   1   2   3   4   5

16

Patterns in directional graphs

Draw the compass direction points in the middle of a large sheet of paper. (Picture 1)

Record the wind direction every day. Draw a square on the correct direction point.

Put the date on the square.

After one week, look at your directional graph.

Which was the most common wind direction?

Draw around all the last squares. (Picture 2)

What does this shape tell you about the wind in your area?

Record the wind direction for a month.

Draw around the last squares.

Is this pattern different?

What does this tell you about the pattern of the winds in your area?

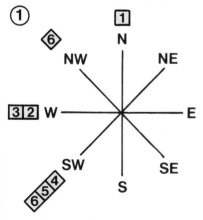

Wind direction
Date . . .

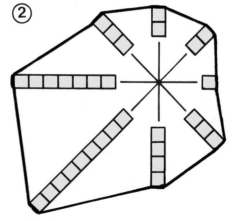

Wind direction for
Date . . .

Comparing directional graphs

Draw twelve radiating lines in the middle of a sheet of paper.
Draw the lines as if they were pointing to the hours on a clock face.
Mark a calendar month on each line instead of the hour. (Picture 1)
Notice that 1 o'clock is January, 2 o'clock is February and so on until 12 o'clock is December.

Go to your library and find the rainfall for each month over the past year in your area (see page 13).
Graph the rainfall on the correct month line. (Picture 1)
Draw a line to link up the rainfall amount for each month. (Picture 2)
What does this tell you?

Repeat this graph for a different year. How is this graph different?

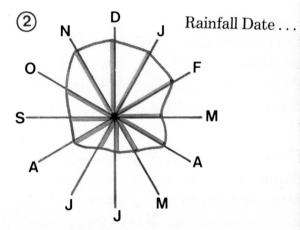

① Rainfall Date . . .

Scale

② Rainfall Date . . .

Record the monthly rainfall for the capital city of another country.
Graph this rainfall as a directional graph. How does their rainfall compare with yours?

Chapter 3 Circle graphs or pie graphs

Clock graphs

Mark out a distance of 100 paces in a straight line.

Borrow a watch that measures seconds.

Ask a friend to time you while you run this 100-pace distance.

How many seconds did it take you?

Draw a clock face on a sheet of paper. (Picture 1)

Mark in the number of seconds that it took you to run 100 paces.

Time your friends as they run the 100-pace distance.

Graph each time as a clock graph. (Picture 2)

Time your friends in other races. Show their times on clock graphs.

Clock graphs can also be used to show hours. Draw a graph to show how many hours you were asleep last night.

① Time taken to run 100 paces

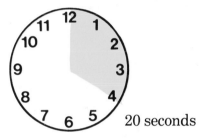

20 seconds

② Time taken to run 100 paces

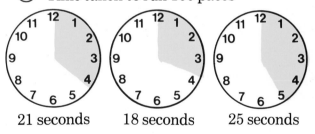

21 seconds   18 seconds   25 seconds

Hours asleep   Date . . .

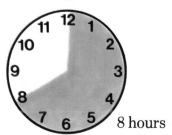

8 hours

Making a percentage protractor

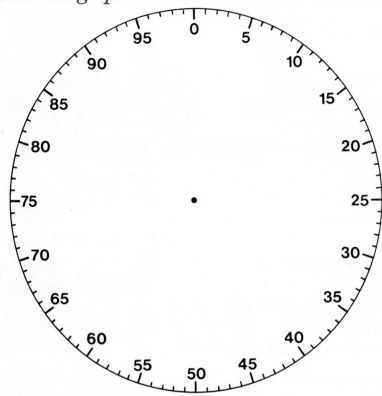

Put a piece of tracing paper onto this drawing of a percentage protractor. Mark in the center of the circle. Carefully mark off the percentage marks.

Use a soft pencil to shade the back of your tracing. (Picture 1)

Place your tracing shaded side down onto a sheet of thin cardboard.

Draw around your tracing again, taking care to mark in each percentage.

Does the protractor drawing come out onto your cardboard?

Ink in the percentage marks.

Carefully cut out your cardboard percentage protractor.

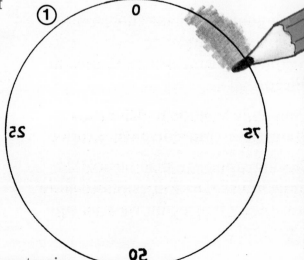

Shade the back of your tracing

20

Making a circle graph

How much allowance do you get
each week?
Carefully record how you spend your
money in one week.
Arrange your spending in groups, such
as food, drinks, clothes, entertainment,
and savings.
How much money did you spend in each
group?

Convert your total for each group into a
percentage of your weekly allowance
(see page 15).
Divide your total for each group by your
weekly allowance.
Multiply each answer by 100. (Picture 1)
Add together each answer. This total
should be 100%.

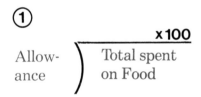

①

Allow-
ance) Total spent
on Food × 100

② How I spent my money

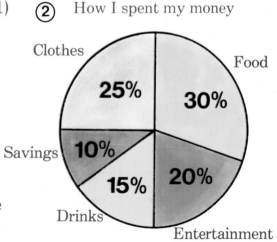

Clothes 25%   Food 30%

Savings 10%   15%   20% Entertainment

Drinks

Place your percentage protractor in the
middle of a sheet of paper (see page 20).
Mark off the percentages for each group.
Draw your circle graph. (Picture 2)
Notice that your circle graph also looks
like slices of a pie.
What takes the biggest slice of your
allowance?

Chapter 4 Line graphs

The space battle game

Draw two large squares on a sheet of paper. Rule 64 squares in each large square. (Picture 1)
Ask a friend to draw 2 similar squares on his sheet of paper.

Mark the vertical columns of squares with numbers. Mark the horizontal columns of squares with letters. (Picture 1) Identify squares B2, E6, and C

Each player has a space fleet of a starship of 5 squares, a battleship of 4 squares, a destroyer of 3 squares, a fighter of 2 squares, and a rocket of 1 square.

Secretly arrange your fleet in one of your large squares of 64 squares. Spaceships can be arranged horizontally or vertically. (Picture 2)

Ask your friend to secretly arrange his fleet in his 64 squares. Take turns firing missiles at the squares. Call out the leter and number to identify each square that your missile hits. If your missile hits a spaceship square your friend must tell you that you have scored a hit. Keep a record of your missile shots on your other square of 64 suares.

The winner is the person who first destroys the other player's spaceship fl

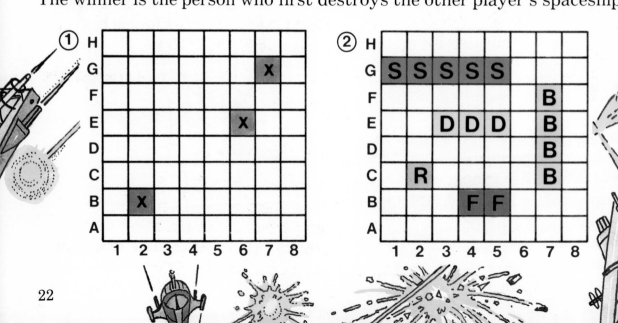

22

Using vectors

If a spaceship moves as in Picture 1, we can show this movement as a vector. First count the horizontal squares that the spaceship moved. Next count the vertical squares that the spaceship moved. The spaceship vectors are $\begin{bmatrix} 3 \\ 2 \end{bmatrix}$.

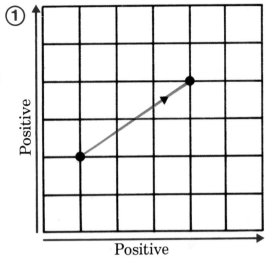

moving in a negative direction. So the vectors are negative. The spaceship A has vectors $\begin{bmatrix} -4 \\ -1 \end{bmatrix}$.

Spaceship B is moving only horizontally, so spaceship B has vectors $\begin{bmatrix} 3 \\ 0 \end{bmatrix}$.

Draw spaceships on a sheet of squared paper. Move each spaceship. Show each movement as an arrow. Calculate the vectors for each movement.

Draw the movement of 4 spaceships. The first moves $\begin{bmatrix} 3 \\ 4 \end{bmatrix}$.
The next moves $\begin{bmatrix} 2 \\ 5 \end{bmatrix}$.
The third moves $\begin{bmatrix} -4 \\ 0 \end{bmatrix}$.
The fourth moves $\begin{bmatrix} -2 \\ -2 \end{bmatrix}$.
A vector describes a journey in length and direction.
Notice it does not tell you where the journey started.

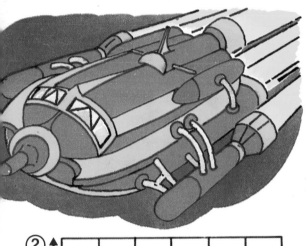

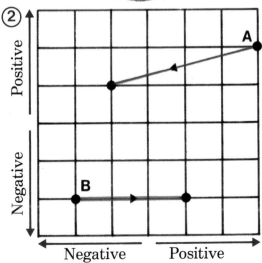

A straight line graph

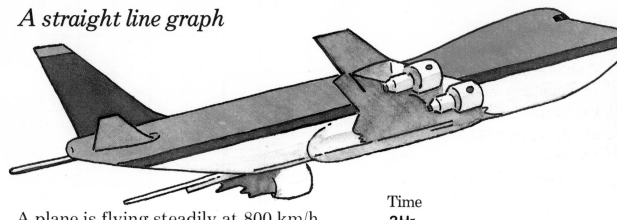

A plane is flying steadily at 800 km/h (497 mph). The vectors are 800 km (497 miles) and 1 hour. Draw a graph of distance and time (Picture 1). If the plane starts at zero distance and time, in one hour it travels 800 km (497 miles) (Picture 1). How far will it travel in 2 hours? Draw a straight line graph of this plane's flight. (Picture 2)

How far will the plane fly in 3 hours? Check your graph for the answer. How far will the plane fly in 2½ hours?

Read your graph to find out how long it takes the plane to fly 2000 km (1242 miles). How long does it take to fly 1000 km (621 miles)?

How long will it take this plane to fly to each of the airports that you drew on your map on page 16?

Draw a graph to show a plane traveling at 500 km/h (311 mph). How much longer does this plane take to fly to each of your airports?

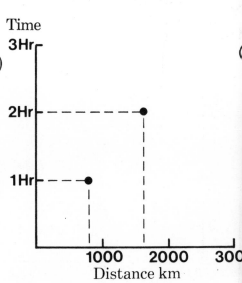

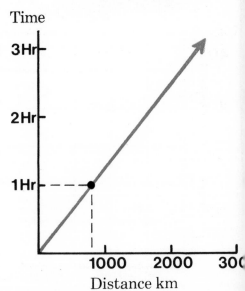

24

Conversion graphs

Travelers visiting different countries will often have to use different units of measurement. Eight km is approximately equal to 5 miles.
How many km are equal to 10 miles? You can use arithmetic to work out the answer, but it is easier to use a graph.

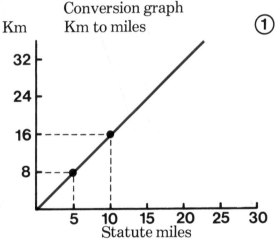

Conversion graph
Km to miles

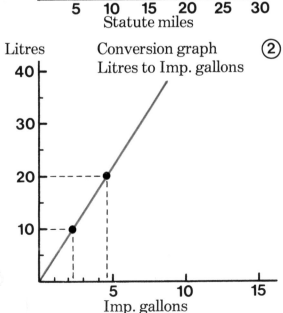

Conversion graph
Litres to Imp. gallons

① Draw a graph showing the relationship between km and miles. (Picture 1)
Use your graph to calculate how many miles are equal to 24 km, 20 km, 12 km, and 18 km.
How many km are equal to 20 miles, 25 miles, 22 miles, and 12 miles?

② Make a conversion graph of litres and imperial gallons. (Picture 2)
Ten litres is equal to 2.2 imperial gallons.
How many imperial gallons are there in 20 litres? Use your graph for other conversions.

Ten litres is equal to 2.6 U.S. gallons. Draw this conversion graph.

Ten inches equals 25.4 cm. Draw a conversion graph of cm and ins.

Look at other conversion tables. Make your own conversion graphs.

Graphs with negative numbers

A thermometer with a Celsius scale is used worldwide, but there are still thermometers with a Fahrenheit scale. (Picture 1)

The Celsius scale has the boiling point of water at 100°C and the freezing point at 0°C.

The Fahrenheit scale has the boiling point of water at 212°F and the freezing point at 32°F.

Draw a conversion graph of Celsius and Fahrenheit. (Picture 2)

Use your graph to convert +40°C, +60°C, and +70°C to Fahrenheit.

Use your graph to convert +50°F, +70°F, +80°F, and +90°F to Celsius.

Notice that temperatures can also fall below zero on both scales.

Convert 0°F, −10°F, and −20°F to Celsius.

Convert −10°C, −15°C, and −20°C to Fahrenheit.

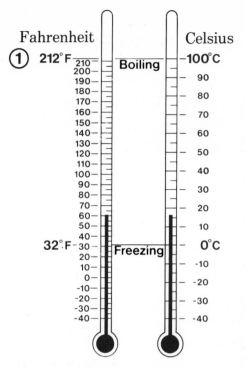

① Fahrenheit / Celsius

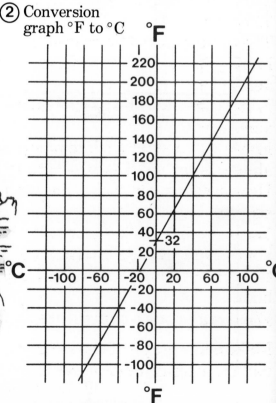

② Conversion graph °F to °C

Making line graphs

Use a thermometer to find the outside shade temperature. Record the temperature every hour on the hour.
What is your highest temperature reading for the day?

Find a sheet of graph paper.
Can you fit the highest temperature onto your graph? (See page 11.)
What scale will you use?
Time should be drawn horizontally on your graph. (Picture 1)
This is called the horizontal axis.
The temperature will be drawn vertically on the vertical axis.
Mark your temperature scale on the vertical axis.
Mark your hours on the horizontal axis. (Picture 1)
Mark with a point each temperature above its correct time. (Picture 1)
Use a ruler to join each point with a straight line.
You have made a line graph. From your graph you can estimate the temperature at any time during each hour.
What was your temperature at 10.30 A.M.?

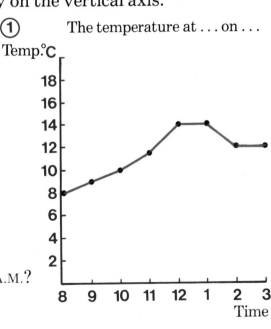

① The temperature at . . . on . . .

Make a line graph of the highest daily temperature for a week.

27

Comparison line graphs

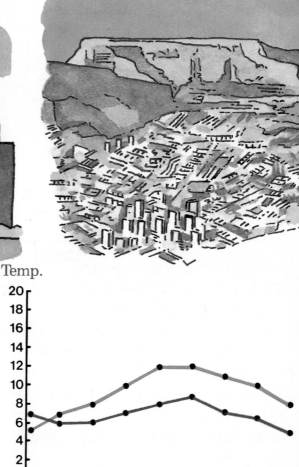

Make a graph of the outside shade temperature each hour on the hour (see page 27).

Record the outside shade temperatures every hour on the hour for a different day.

Graph these new temperatures on your old temperature graph.

Use a different color to draw the graph of new temperatures.

What can you tell from your comparison line graph?

Temp.

Date ——— Date ———

Go to your library. Make comparison graphs showing the average monthly temperatures for each month of the year for different cities in the world. What can you tell from these graphs?

Make comparison line graphs showing your marks in different school subjects over a period of time.

Make comparison line graphs showing the scores for different teams.

Try to invent your own comparison line graphs.

Drawing a profile of uneven ground

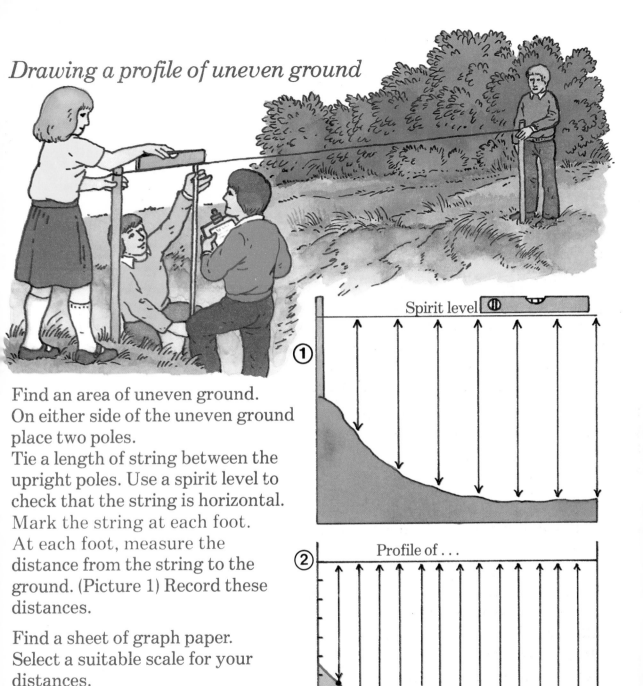

Spirit level

①

Profile of . . .

②

Find an area of uneven ground.
On either side of the uneven ground
place two poles.
Tie a length of string between the
upright poles. Use a spirit level to
check that the string is horizontal.
Mark the string at each foot.
At each foot, measure the
distance from the string to the
ground. (Picture 1) Record these
distances.

Find a sheet of graph paper.
Select a suitable scale for your
distances.
Plot the distances from the string to
the ground on your graph.
Join these points and you have made
a profile of the ground. (Picture 2)

Curved graphs

Compare addition with multiplication.
Here are some addition facts:

$$
\begin{array}{cccccccccc}
1 & 2 & 3 & 4 & 5 & 6 & 7 & 8 & 9 & 10 \\
+ & + & + & + & + & + & + & + & + & + \\
1 & 2 & 3 & 4 & 5 & 6 & 7 & 8 & 9 & 10 \\
\hline
2 & 4 & 6 & 8 & 10 & 12 & 14 & 16 & 18 & 20
\end{array}
$$

Graph the numbers against their sums. (Picture 1)
Notice that you get a straight line as
the numbers increase equally.

Here are some multiplication facts:

$$
\begin{array}{cccccccccc}
1 & 2 & 3 & 4 & 5 & 6 & 7 & 8 & 9 & 10 \\
\times & \times & \times & \times & \times & \times & \times & \times & \times & \times \\
1 & 2 & 3 & 4 & 5 & 6 & 7 & 8 & 9 & 10 \\
\hline
1 & 4 & 9 & 16 & 25 & 36 & 49 & 64 & 81 & 100
\end{array}
$$

Graph these numbers against their
products. (Picture 2)
What do you notice about how the
product increases?

Draw a square with sides of 4 inches.
The perimeter is the distance around
the square.
So the perimeter of your square is
4 inches $\times$ 4 inches = 16 inches.
The area is the length times the
width, which is 4 inches $\times$ 4 inches
= 16 square inches.
Draw a graph showing the
relationship between the perimeters
and the areas of squares with sides
of 2 inches, 4 inches, 6 inches, and
8 inches.

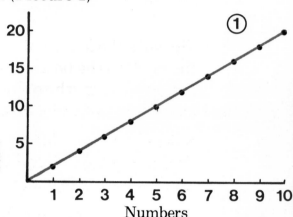

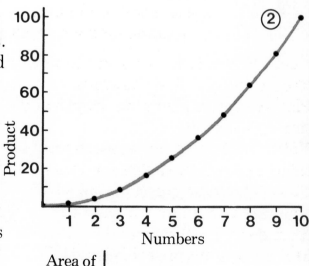

30

Timing a pendulum

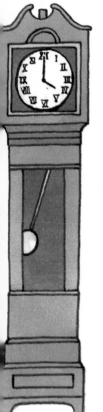

Make a hook from a wire paper clip. (Picture 1)
Tape the hook to the middle of a doorway arch. (Picture 2)
Fix a small ball of modeling clay to
the end of a length of thread.
You will be shortening your
pendulum, so pass the pendulum thread
through the hook and tape the extra
thread to the door frame. (Picture 2)

Measure the length of thread from
the hook to the bottom of the clay bob.
Adjust the length so that it is an
exact number of inches.

Release your pendulum. Record the
time of release. Record the time
taken for 40 swings of your pendulum.

Shorten the pendulum by 5 inches.
Release the pendulum. Record the
time taken for another 40 swings.

Keep shortening your pendulum by 5 inches.
Record the time for 40 swings.
Plot a graph of pendulum length and
time. (Picture 3)
Notice that if you use a line to join
each of your results you produce a
curve. Use your graph to find what
length of pendulum will take 40
seconds to do 40 swings.
Make a pendulum of this length.
Does this pendulum take one second to do one swing?
You have made a second timer.

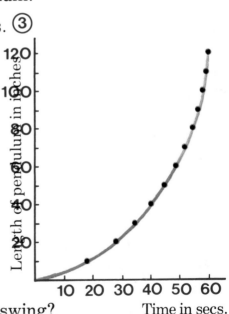

Hook

Tape

Modeling clay

Length of pendulum in inches

120
100
80
60
40
20

10 20 30 40 50 60
Time in secs.

31

String art

Graphs can give you some very beautiful designs.

Draw the vertical and horizontal axes of a graph on a sheet of paper.

Number each axis equally, reversing the numbers on the vertical axis. (Picture 1)

Use a ruler to join together all the identical numbers. Join 1 to 1, 2 to 2 etc. What happens?

Repeat this. This time make one set of numbers farther apart. (Picture 2)

Repeat this, but this time alter the angle between the axes. (Pictures 3 and 4)

Repeat this. This time do not connect the axes. (Picture 5)

Try different shapes and angles. Draw your results. Try larger shapes.

Draw one of your shapes onto cardboard. Put holes in each number.
Join the numbers with colored thread by passing the thread through the holes.

Go to your library and find a book on curve stitching or string art.

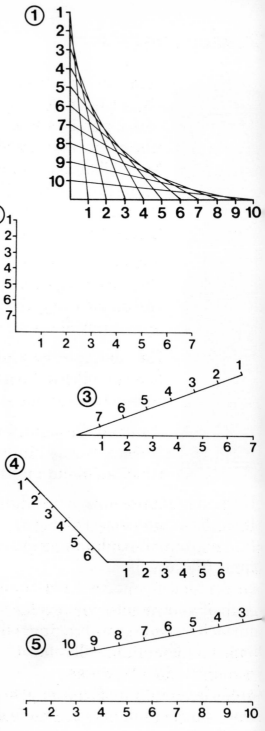